Open architecture controll

Sakib Sarguroh
Arun Rane

Open architecture controller to operate a CNC machine

Utilizing GRBL-Arduino-based controller to operate a computerized numerical control machine

LAP LAMBERT Academic Publishing

Imprint

Cover image: www.ingimage.com

Publisher:
LAP LAMBERT Academic Publishing
is a trademark of
International Book Market Service Ltd., member of OmniScriptum Publishing Group
17 Meldrum Street, Beau Bassin 71504, Mauritius
Printed at: see last page
ISBN: 978-620-2-51429-3

Abstract

This book presents a method for eliminating the proprietary control for computerized numerical control (CNC) machine with aid of a controller which will take G and M codes generated by computer-aided design/computer aided manufacturing (CAD/CAM) software for a two dimensional drawing file (.prt or .dxf) as input. The drawing file is obtained by converting a JPEG or PNG image into .prt or .dxf file with the aid of Img2CAD software. This drawing file is imported in UG NX CAD/CAM software (Version 5) and scaled to the required dimensions. 2D manufacturing of the drawing is carried out in UG Nx and G and M codes for the same is obtained by carrying out post-processing operation. G and M code program is saved in a text file with extension .nc. This file is taken as an input by Universal Gcode Sender software which is connected to Grbl shield V5 and Arduino Uno hardware via serial port of laptop. Before executing G code program, the output can be visualized using gcode visualizer option. G and M code file will then be executed through Universal Gcode Sender software to run two stepper motors which will move two mechanical systems(x and y axis) through timing belt and pulley to draw the desired image. Accuracy and performance of the system will be tested by drawing a rectangle, hexagon or circle and measuring their dimension.

Table of contents

List of figures

List of tables

List of abbreviations

CNC	Computerized numerical control
CAD	Computer aided designing
CAM	Computer aided manufacturing
JPEG	Joint photographic experts group
PNG	Portable network graphics
UG NX	Unigraphics Next generation
UGS	Universal GcodeSender

Chapter 1

Introduction

1.1. Motivation

CNC machines play a vital role in manufacturing industry and it has become challenging to create more agile and adaptable solutions. Majority of the operations like milling, cutting, drilling are performed by CNC machines without having the ability to control and enhance inputs. CNC machines require heavy initial investment and its proprietary control prevents modification to the program codes which is generally retained by vendors who distribute the software in compiled form. By eliminating proprietary control, it will enable provision for customization allowing users with controls to boost performance. Continuous improvement can be achieved which will greatly increase investment returns. Windows based CNC control provides better traceability, scalability, connectivity and versatility enabling cost reduction and quality improvement.

CNC machines utilize G and M code language generated by Computer Aided Manufacturing Systems (CAM) which uses Computer Aided Design (CAD) data. G and M codes language is characterized by numerical codes such as G, M, F, T, S etc illustrating the operations of a machine. G and M codes generated by controllers of various vendors are not universal.

This book aims at presenting a method of utilizing a generalized or universal controller which will take G and M codes generated by any CAD/CAM software. The controller will then convert these codes into step pulses which will be sent as an input to actuator or stepper motor of CNC machine.

1.2. Problem Statement

The problem can be stated as generalization of controller which can take G and M codes generated by any CAD/CAM software and execute it by means of a windows based application to draw the desired figure using simple hardware setup.

1.3. Scope and Objectives

Following objectives are set for the project

1) To develop a universal controller; which will take G and M codes generated by any CAD/CAM software.
2) To convert JPEG or PNG image of a rectangle or circle or polygon into drawing file (.dxf or .prt) and generate gcode file for the same after importing it in UG NX 5.0.
3) To develop a simple two axis CNC machine which will incorporate two 2.6kg/cm stepper motors which will draw a rectangle or circle or polygon after getting input of gcode file of rectangle or circle or polygon through universal gcode sender software via Arduino Uno and Grbl shield V5 hardware.
4) To measure the accuracy of the drawn rectangle or circle or polygon.

1.4. Research Methodology

The block diagram in figure 1.4.1 describes the research methodology used for current work.

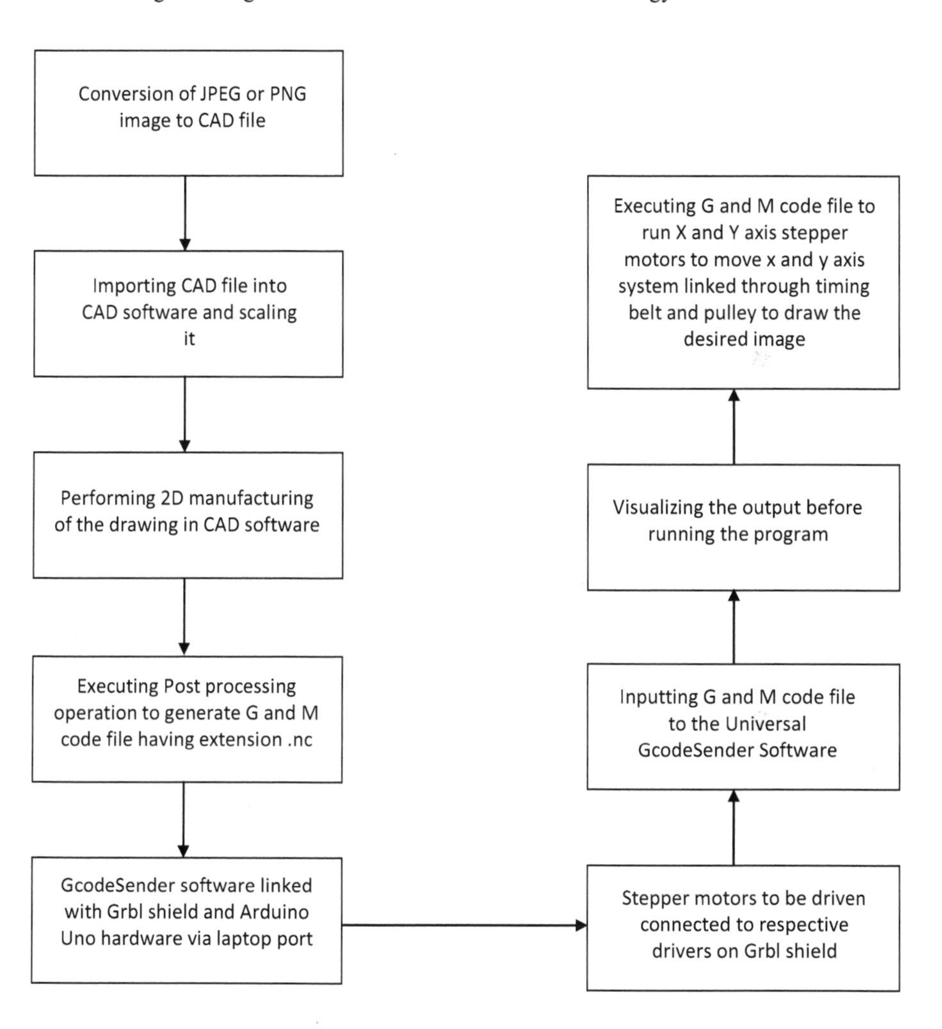

Figure 1.4.1: Block diagram of Research Methodology

1.5. Organization of the project

Chapter 2 gives an insight into background of various components forming the controller. Detailed information on how the controller has been designed is discussed in chapter 3. In chapter 4 complete processes on how gcode file is generated in UG NX 5 CAD/CAM software is explained. Chapter 5 gives explanation on various hardware components that have been used to form the complete assembly and gcode program execution to draw the desired image is also explained.

Chapter 2

Literature Review

Banzi et al. (2005) along with other co-founders started Arduino as a project at the Interaction Design Institute Ivrea, Italy. Arduino is an open source platform with easy to use hardware and software. Arduino boards can read inputs such as light sensing and convert into an output by activating a motor or turning on an LED. Banzi's et al. (2005) Uno shown in figure 2.1 was the first in the series of Arduino boards.

Figure 2.1: Arduino Uno

Banzi's et al. (2005) Arduino Uno hardware is an open source design and all its components are off the shelf. At the heart of the Arduino is microcontroller chip which can be programmed to do various operations. It also has timing crystal, power regulator, USB interface and power jack. It has 14 digital I/O pins out of which 6 provide PWM output. Uno Board also has 6 analog input pins along with power connection pins for 3.3V and 5V supply.

Banzi's et al. (2005) Arduino Uno software shown in figure 2.2 is an open source in which desired code can be written and uploaded on the Uno board. It runs on Windows, Mac OS X and Linux.

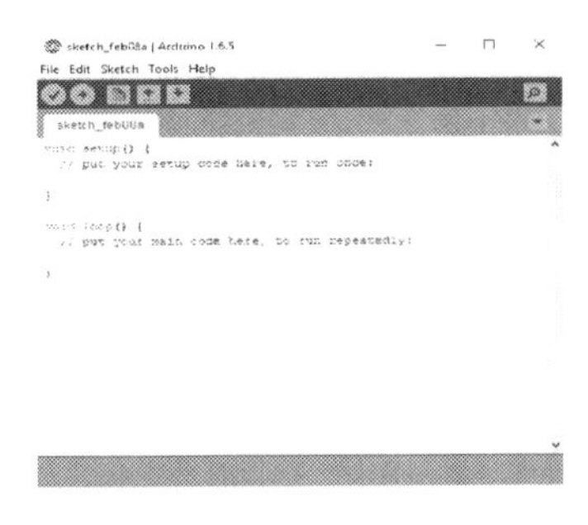

Figure 2.2: Arduino software (IDE)

Skogsrud (2009) graced the open-source community by releasing early versions of Grbl in 2009. Grbl is moving ahead under the new leadership of Jeon (2011) since 2011. Grbl shown in figure 2.3 is a high performance, low cost, open source different from parallel port based motion management for CNC machine. It's written in optimized C which is compatible with and runs on Arduino Uno. The controller is written in extremely optimized C utilizing each clever feature of the AVR-chips to attain precise timing and asynchronous operation. It's ready to maintain up to 30 kHz of stable, disturbance free control pulses.

Figure 2.3: Grbl Shield V5

Grbl is for three axis machines. It accepts standards-compliant g-code and has been tested with the output of many CAM tools with no issues. Grbl interprets a subset of RS274NGC standard gcode. Arcs, circles and helical motion are completely supported, as well as, all different primary g-code commands. Grbl includes full acceleration management with look ahead which means the controller will look up to 18 motions into the future and set up its velocities ahead to deliver smooth acceleration and jerk-free cornering. Grbl comprises of 12 V supply connection along with three drivers for stepper motors. It also has 3 trimpots for each axis to control amount of current flow. Micro stepping provision up to 8x is also present on the Grbl.

Table 2.1 shows the list of few G and M codes that are utilized by the controller.

Table 2.1: Gcodes utilized by Grbl

Modal Group Meaning	Member Words
Motion Mode	G0, G1, G2, G3, G38.2, G38.3, G38.4, G38.5, G80
Coordinate System Select	G54, G55, G56, G57, G58, G59
Plane Select	G17, G18, G19
Distance Mode	G90, G91
Arc IJK Distance Mode	G91.1
Feed Rate Mode	G93, G94
Units Mode	G20, G21
Cutter Radius Compensation	G40
Tool Length Offset	G43.1, G49
Program Mode	M0, M1, M2, M30
Spindle State	M3, M4, M5
Coolant State	M7, M8, M9

Saunders (2014) has demonstrated the running of Nema stepper motors with the aid of Arduino Uno-Grbl shield as main hardware and Universal Gcode sender software via computer.

Chapter 3

Design of controller

3.1. Grbl shield and Arduino Uno connection with stepper motor

Arduino Uno is a microcontroller board based on the ATmega328P. It has 14 digital input/output pins, 6 analog inputs, a16 MHz quartz crystal, a USB connection, a power jack, an ICSP header and a reset button.

Grbl is an open source, embedded, high performance, low cost gcode parser and CNC milling controller written in optimized C that runs on an Arduino Uno Atmega 328. The controller utilizes every clever feature of AVR chips to achieve precise timing and asynchronous operation.

Figure 3.1.1: Arduino Uno and Grbl Shield V5 in individual and combined state

3.1.1. Connecting power and power supply requirements

Grbl shield uses Arduino's power supply for the digital logic and it requires only motor power to be provided. Arduino can be powered up from its DC input connector, but more commonly it is powered from the USB port of laptop or computer as shown in figure 3.1.1.1.

Figure 3.1.1.1: USB connection of controller with laptop

Grbl shield has 12V connection to which power is supplied through SMPS as shown in figure 3.1.1.2. The care has to be taken to make sure that positive and common terminals of SMPS are connected with positive and common of Grbl shield, if connected otherwise, it will damage the Grbl Shield. In this project, two Nema 17 2.6 kg-cm stepper motors are used for respective x and y axis which require 1-2 amps of current. The stepper drivers can source up to 2.5 amps (with cooling). The motors do not draw maximum power at the same time and it's adequate to provide 2/3 of the total rated current. For two 1.2 amp motors used, minimum 1.6 amps of current is fine. A minimum 4.5 amp supply is sufficient for almost all Nema 17 stepper motors. Motor voltage can vary between +12VDC to +30VDC. The SMPS used in the project provides 15amp, 12VDC supply to the Grbl for running motors which is sufficient as per the requirement discussed above. The motor current for each axis can be adjusted with the 5mm trimpot next to that axis by rotating it in clockwise direction to increase the current and in counter-clockwise direction to decrease it.

Figure 3.1.1.2: SMPS and its connection with Grbl Shield

3.1.2. Stepper motor connection to Grbl shield

The 12V power supply is made use to run stepper motors via inbuilt stepper drivers on Grbl shield. The motors used are bipolar having 4 wires i.e. 2 pairs. Pair of wires can be found out by touching two test leads of ohmmeter to two wires at a time. Most NEMA 17 motors show a pair resistance from 2 to 20 ohms. There is a shortcut to find wire pairs of bipolar motors. Initially, none of the wires end should be touching each other and stepper motor shaft is made to spin with fingers. There will be no resistance and motor will spin freely with ease. As the end wires of a pair are connected, there will be a resistance and it will much harder to spin the motor. Once the pairs have been identified they are fitted to the divers on the Grbl shield as shown in figure 3.1.2.1.

Figure 3.1.2.1: Stepper motor connection to drivers on Grbl shield

3.1.3. Microstepping in Grbl

There are four sets of microstepping available in Grbl – 1x, 2x, 4x and 8x as shown in figure 3.1.3.1. By default microstepping setting is 8x i.e. whatever the dimensions have been taken in UG Nx 5 to generate the gcode file, the drawing obtained will be 1/8th of the original dimensions. In order to change the microstepping setting to 1x or 2x or 4x, we need to add or remove jumpers to 4 pin male headers as shown in figure 3.1.3.1.

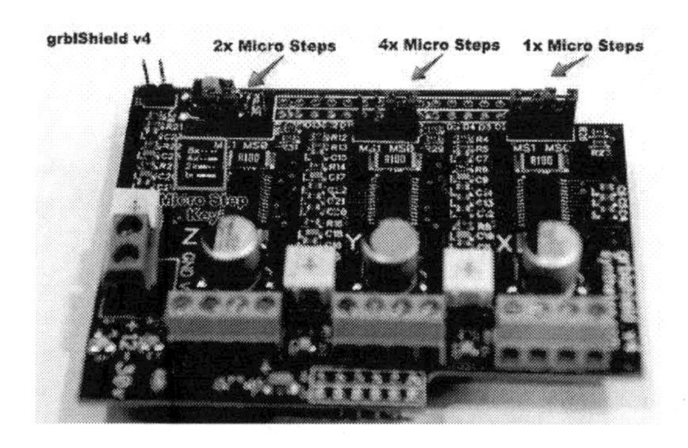

Figure 3.1.3.1: Microstepping in Grbl

3.2. XLoader software

XLoader is software that is used to upload files to Arduino board using the boot loader. It can be downloaded from website http://russemotto.com/xloader/. Figure 3.2.1 shows the screenshot of the XLoader application.

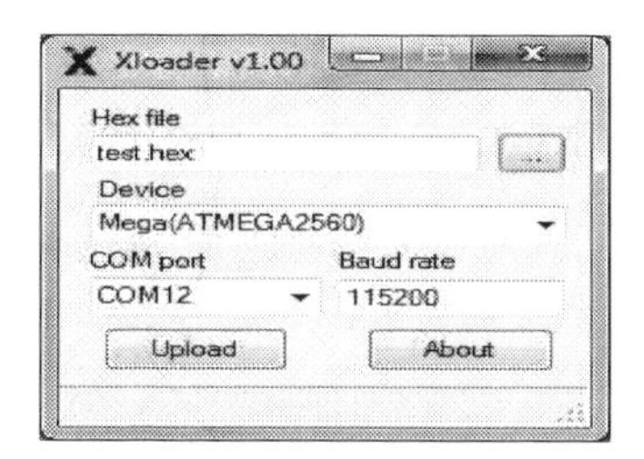

Figure 3.2.1: XLoader application

The desired hex file to be uploaded in Arduino Uno can be browsed and appropriate Arduino must be selected under device option, in our case it is Uno (ATmega). Baud rate indicates number of pulses that are transmitted per unit time. In our case the baud rate used is 9600. After clicking on the upload button hex file is successfully uploaded in Arduino Uno. The desired hex file to be uploaded in Arduino Uno can be downloaded from website https://github.com/grbl/grbl/downloads.

3.3. Universal GcodeSender

Universal GcodeSender is a Java based GRBL compatible cross platform used for interfacing with advanced CNC controllers like GRBL. It is tested on Windows OSX, Linux and Rasberry Pi. It is a self-contained Java application which includes all external dependencies for all supported operating systems, meaning if one has the Java runtime environment setup, UGS provides the rest. Universal GcodeSender can be downloaded from website https://github.com/winder/Universal-G-Code-Sender.

The gcode file generated after CAM operation (discussed in Chapter 4) is given as an input to the Universal GcodeSender software which is linked to the Arduino Uno and Grbl shield V5 hardware via serial communication i.e. Universal Serial Bus(USB) port of a laptop. This section will give an insight into various functionalities available with UGS.

3.3.1. Commands

After clicking on the UGS application, the sub-options under command functionality appear as shown in figure 3.3.1.1. The appropriate port to which the Grbl hardware is connected via laptop port is to be selected. Then baud rate of 9600 is chosen and Grbl as firmware is selected. After clicking on open button, UGS gets connected with Grbl shield which is indicated in the white console box. Typing in any Gcode such as G01 X200 in text box under command option will cause the x axis motor to move through a distance of 200mm, just like any CNC machine.

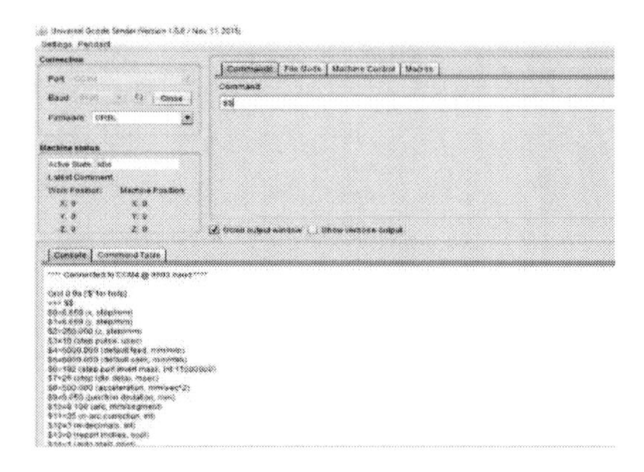

Figure 3.3.1.1: Command Functionality

Once connection is done Grbl-prompt appears as Grbl 0.9i ['$' for help]. There are various settings available in Grbl which can be viewed and modified through command functionality. $ Symbol can be typed in and after hitting the enter button Grbl prints a help message showing other settings which can be used as shown in table 3.3.1.1.

Table 3.3.1.1: Grbl overall settings

Symbol	Function
$$	view Grbl settings
$#	view # parameters
$G	view parser state
$I	view build info
$N	view startup blocks
$x=value	save Grbl setting
$Nx=line	save startup block
$C	check gcode mode
$X	kill alarm lock
$H	run homing cycle
~	cycle start
!	feed hold
?	current status
ctrl-x	reset Grbl

The $-commands are Grbl system commands used to tweak the settings, view or change Grbl's states and running modes, and start a homing cycle. The last four non-$ commands are real time control commands that can be sent at anytime, no matter what Grbl is doing. These either immediately change Grbl's running behavior or immediately print a book of the important real time data like current position.

Grbl settings

$$ - View Grbl settings

To view the settings, $$ is typed and enter button is pressed after connecting to Grbl. Grbl responds with a list of the current system settings, as shown in the table 3.3.1.2. All of these settings are persistent and kept in EEPROM, so if power supply is cut off, these will be loaded back up the next time Arduino is powered up.

Table 3.3.1.2: Grbl $$ settings

Symbol	Terminology	Default value	Unit
$0	Travel across x-axis	250	step/mm
$1	Travel across y-axis	250	step/mm
$2	Travel across z-axis	250	step/mm
$3	Step pulse	10	usec
$4	Default feed	250	mm/min
$5	Default seek	500	mm/min
$6	Step port invert mask	192	
$7	Step idle delay	25	msec
$8	Acceleration	10	mm/sec^2
$9	Junction deviation	0.05	mm
$10	Arc	0.1	mm/segment
$11	n-arc correction	25	int
$12	n-decimals	3	int
$13	Book inches	0	bool
$14	Auto start	1	bool
$15	Invert step enable	0	bool

$16	Hard limits	0	bool
$17	Homing cycle	0	bool
$18	Homing dir invert mask	0	int
$19	Homing feed	25	mm/min
$20	Homing seek	250	mm/min
$21	Homing debounce	100	msec
$22	Homing pull-off	1	mm

$0, $1 and $2 – [X, Y, Z] steps/mm

Grbl needs to know how far each step will take the tool in reality. To calculate steps/mm for an axis of the machine the following parameters must be known:

- The mm traveled per revolution of the stepper motor. This is dependent on the belt drive gears or lead screw pitch.
- The full steps per revolution of the steppers (typically 200).
- The microsteps per step of the controller (typically 1, 2, 4, 8, or 16). Using high microstep values (e.g., 16) can reduce the stepper motor torque, so the lowest that gives the desired axis resolution and comfortable running properties must be used.

The steps/mm calculation has been done at the end of this section. This value has to be computed for every axis that will be put to use.

$3 – Step pulse, microseconds

Stepper drivers are rated for a certain minimum step pulse length. The shortest pulses the stepper drivers can reliably recognize must be used. If the pulses are too long, it can cause trouble when running the system at very high feed and pulse rates, because the step pulses can begin to overlap each other. 10 microseconds is recommended, which is the default value.

$6 – Step port invert mask:binary

This setting inverts the step pulse signal. By default, a step signal starts at normal-low and goes high upon a step pulse event. After a step pulse time set by $3, the pin resets to low, until the next step pulse event. When inverted, the step pulse behavior switches from normal-high, to low during the pulse, and back to high. This setting is generally not required, but this can be useful for certain CNC-stepper drivers that have peculiar requirements. For example, an artificial delay between the direction pin and step pulse can be created by inverting the step pin.

This invert mask setting is a value which stores the axes to invert as bit flags. The settings value needs to be entered for the axes that are required to be inverted. For example, to invert the X and Z axes, we need to input $6=5 to Grbl and the setting will now read $6=5 (step port invert mask:00000101). Various invert combinations along three axes have been shown in table 3.3.1.3.

Table 3.3.1.3: Combinations for step port invert

Setting Value	Mask	Invert X	Invert Y	Invert Z
0	0	N	N	N
1	1	Y	N	N
2	10	N	Y	N
3	11	Y	Y	N
4	100	N	N	Y
5	101	Y	N	Y
6	110	N	Y	Y
7	111	Y	Y	Y

$7 - Step idle delay, msec

Every time steppers complete a motion and come to a stop, Grbl delays disabling the steppers by this value or, the axes can be kept enabled (powered so as to hold position) by setting this value to the maximum 25 milliseconds i.e. $7=25.

The stepper idle lock time is the time length Grbl will keep the steppers locked before disabling. Depending on the system, this can be set to zero and disable it. On others, 25-50 milliseconds is needed to make sure the axes come to a complete stop before disabling. This is to help account for machine motors that do not like to be left on for long periods of time without doing something. Also, some stepper drivers don't remember which micro step they stopped on, so when re-enabled, some lost steps may be witnessed due to this. In this case, the steppers need to be kept enabled via $7=25.

$8 – [X, Y, Z] Acceleration, mm/sec^2

This sets the axes acceleration parameters in mm/second/second. Simplistically, a lower value makes Grbl ease slower into motion, while a higher value yields tighter moves and reaches the desired feed rates much quicker. The simplest way to determine the values for this setting is to individually test each axis with slowly increasing values until the motor stalls. Then acceleration setting needs to be finalized with a value 10-20% below this absolute max value. This should account for wear, friction, and mass inertia.

$9 - Junction deviation, mm

Junction deviation is used by the acceleration manager to determine how fast it can move through line segment junctions of a G-code program path. For example, if the G-code path has a sharp 10 degree turn coming up and the machine is moving at full speed, this setting helps determine how much the machine needs to slow down to safely go through the corner without losing steps.

The higher value gives faster motion through corners, while increasing the risk of losing steps and positioning. Lower value makes the acceleration manager more careful and will lead to careful and slower cornering. If machine tries to take a corner too fast, this value needs to be decreased to make it slow down when entering corners. If machine needs to be moved faster through junctions, this value needs to be increased to speed it up.

$10 – Arc tolerance, mm

Grbl renders G2/G3 circles, arcs, and helices by subdividing them into tiny lines, such that the arc tracing accuracy is never below this value. The default value is 0.1mm and is well below the accuracy of most CNC machines. If the circles are too crude or arc tracing is performing slowly, this setting needs to be adjusted. Lower values give higher precision but may lead to performance issues by overloading Grbl with too many tiny lines. Alternately, higher values traces to a lower precision, but can speed up arc performance since Grbl has fewer lines to deal with.

Arc tolerance is defined as the maximum perpendicular distance from a line segment with its end points lying on the arc, aka a chord. With some basic geometry, one can solve for the length of the line segments to trace the arc that satisfies this setting. Modeling arcs in this way is great, because the arc line segments automatically adjust and scale with length to ensure optimum arc tracing performance, while never losing accuracy.

$13 - Book inches, bool

Grbl has a real-time positioning booking feature to provide a user feedback on where the machine is exactly at that time, as well as, parameters for coordinate offsets and probing. By default, it is set to book in mm, but by sending a $13=1 command, one can send this boolean flag to true and these booking features will now book in inches. $13=0 can be input to set back to mm.

$15 - Step enable invert, bool

By default, the stepper enable pin is high to disable and low to enable. If the setup needs the opposite, the stepper enable pin needs to be inverted by typing $15=1. It can be disabled by giving input as $15=0.

$16 - Hard limits, bool

Hard limits use physical switches. Basically when some switches are wired up (mechanical, magnetic, or optical) near the end of travel of each axes, or where there might be trouble if the program moves too far to where it shouldn't. Then the switch triggers and it will immediately halt all motion, shutdown the coolant and spindle (if connected), and go into alarm mode, which forces one to check the machine and reset everything.

To use hard limits with Grbl, the limit pins are held high with an internal pull-up resistor, so all that has to done is wire in a normally-open switch with the pin and ground and enable hard limits with $16=1(disable it with $16=0). If a limit has to be kept at both ends of travel of one axes, two switches need to be wired in parallel with the pin and ground, so if either one of them trips, it triggers the hard limit.

A hard limit event is considered to be critical event, where steppers immediately stop and will likely have lost steps. Grbl doesn't have any feedback on position, so it can't guarantee it has any idea where it is. So, if a hard limit is triggered, Grbl will go into an infinite loop ALARM mode, giving a chance to check the machine and forcing to reset Grbl. It's purely a safety feature.

$17 - Homing cycle, bool

The homing cycle is used to accurately and precisely locate a known and consistent position on a machine every time Grbl starts up between sessions. In other words, one can exactly know where the tool position is at any given time, every time. For example, while machining something or are about to start the next step in a job and the power goes out, and Grbl is restarted and Grbl has no idea where it is. One is left with the task of figuring out where the machine is. With homing, one always has the machine zero reference point to locate from, so all that has to be done is run the homing cycle and resume where the machine left off.

To set up the homing cycle for Grbl, one needs to have limit switches in a fixed position that won't get bumped or moved, or else the reference point gets messed up. Usually they are setup in the farthest point in +x, +y, +z of each axis. The limit switches need to be wired in with the limit pins and ground, just like with the hard limits, and enable homing. By default, Grbl's homing cycle moves the Z-axis positive first to clear the workspace and then moves both the X and Y-axes at the same time in the positive direction.

Also, when homing is enabled, Grbl will lock out all G-code commands until the homing cycle is performed i.e. no axis motions will happen unless the lock is disabled ($X). Most, if not all CNC controllers, do something similar, as it is mostly a safety feature to prevent users from making a positioning mistake, which is very easy to do.

$18 - Homing dir invert mask, int:binary

By default, Grbl assumes the homing limit switches are in the positive direction, first moving the z-axis positive, then the x-y axes positive before trying to precisely locate machine zero by going back and forth slowly around the switch. If the machine has a limit switch in the negative direction, the homing direction mask can invert the axes' direction. It works just like the step port invert and direction port invert masks, where all that has to be done is send the value in the table to indicate which axis is required to be inverted and searched for in the opposite direction.

$19 - Homing feed, mm/min

The homing cycle first searches for the limit switches at a higher seek rate, and after it finds them, it moves at a slower feed rate to home into the precise location of machine zero. Homing feed rate is that slower feed rate. This rate can be set to value that provides repeatable and precise machine zero locating.

$20 - Homing seek, mm/min

Homing seek rate is the homing cycle search rate, or the rate at which it first tries to find the limit switches. Whatever rate at which it gets to the limit switches is adjusted in a short enough time without crashing into the limit switches if they come in too fast.

$21 - Homing debounce, ms

Whenever a switch triggers, some of them can have electrical/mechanical noise that actually bounce the signal high and low for a few milliseconds before settling in. To solve this, the signal needs to be debounce, either by hardware with some kind of signal conditioner or by software with a short delay to let the signal finish bouncing. Grbl performs a short delay, only homing when locating machine zero. This delay is set to value to whatever the switch needs to get repeatable homing. In most cases, 5-25 milliseconds is the value set.

$22 - Homing pull-off, mm

To play nice with the hard limits feature, where homing can share the same limit switches, the homing cycle will move off all of the limit switches by this pull-off travel after it completes. In other words, it helps to prevent accidental triggering of the hard limit after a homing cycle.

$x=val - Save Grbl setting

The $x=val command saves or alters a Grbl setting, which can be done manually by sending this command when connected to Grbl through a serial terminal program, but most Grbl GUIs will do this automatically as a user-friendly feature.

To manually change e.g. the microseconds step pulse option to 10us the following has to be typed in and enter has to be pressed.

$3=10

If everything went well, Grbl will respond with an ok and this setting is stored in EEPROM and will be retained forever or until it has been changed again. It can be checked if Grbl has received and stored the setting correctly by typing $$ to view the system settings again.

Grbl's other $ commands:

The other $ commands provide additional controls for the user, such as printing feedback on the current G-code parser modal state or running the homing cycle. This section explains what these commands are and how to use them.

$# - View gcode parameters

G-code parameters store the coordinate offset values for G54-G59 work coordinates, G28/G30 pre-defined positions, G92 coordinate offset, tool length offsets, and probing. Most of these parameters are directly written to EEPROM anytime they are changed and are persistent i.e. they will remain the same, regardless of power-down, until they are explicitly changed. The non-persistent parameters, which are not retained when reset or power-cycled, are G92, G43.1 tool length offsets, and the G38.2 probing data.

G54-G59 work coordinates can be changed via the G10 L2 Px or G10 L20 Px command defined by the NIST gcode standard and the EMC2 (linuxcnc.org) standard. G28/G30 pre-defined positions can be changed via the G28.1 and the G30.1 commands, respectively.

When $# is called, Grbl will respond with the stored offsets from machine coordinates for each system as shown in table 3.3.1.4. TLO denotes tool length offset, and PRB denotes the coordinates of the last probing cycle.

Table 3.3.1.4: Gcode parameter settings

System	Stored offset (X,Y,Z)
G54	4.000,0.000,0.000
G55	4.000,6.000,7.000
G56	0.000,0.000,0.000
G57	0.000,0.000,0.000
G58	0.000,0.000,0.000
G59	0.000,0.000,0.000
G28	1.000,2.000,0.000
G30	4.000,6.000,0.000
G92	0.000,0.000,0.000
TLO	0.000,0.000,0.000
PRB	0.000,0.000,0.000

$G - View gcode parser state

This command prints all of the active gcode modes in Grbl's G-code parser. When sending this command to Grbl, it will reply with something like:
[G0 G54 G17 G21 G90 G94 M0 M5 M9 T0 S0.0 F500.0]

These active modes determine how the next G-code block or command will be interpreted by Grbl's G-code parser. Modes set the parser into a particular state without having to constantly tell the parser how to parse it. These modes are organized into sets called modal groups that cannot be logically active at the same time. For example, the units modal group sets whether the G-code program is interpreted in inches or in millimeters. In addition to the G-code parser modes, Grbl will book the active T tool number, S spindle speed, and F feed rate, which all default to 0 upon a reset.

$I - View build info

This prints feedback to the user, the Grbl version and source code build date. Optionally, $I can also store a short string to help identify which CNC machine it is communicating with, if more than one machine is using Grbl. To set this string, Grbl $I=xxx can be sent, where xxx is the customization string that is less than 80 characters. The next time a query is sent to Grbl with a $I view build info; Grbl will print this string after the version and build date.

$N - View startup blocks

$Nx are the startup blocks that Grbl runs every time Grbl is powered ON or reset. In other words, a startup block is a line of G-code that can have Grbl auto-magically run to set the G-code modal defaults, or anything else that makes Grbl to do every time the machine is started. Grbl can store two blocks of G-code as a system default. So, when connected to Grbl, $N has to be typed in and enter button pressed. Grbl responds with something short like:
$N0=
$N1=
ok

This just means that there is no G-code block stored in line **$N0** for Grbl to run upon startup. **$N1** is the next line to be run.

$Nx=line - Save startup block

To set a startup block, **$N0=** followed by a valid G-code block has to be entered and enter button pressed. Grbl runs the block to check if it's valid and then replies with an **ok** or an error: to tell if it's successful or something went wrong. If there is an error, Grbl will not save it.

For example, the first startup block `$N0` to be set as the G-code parser modes like G54 work coordinate, G20 inches mode, G17 XY-plane; $N0=G20 G54 G17 has to be typed in and enter pressed; and an ok response can be seen. It can be checked if it got stored by typing $N and a response like $N0=G20G54G17 will appear. Once a startup block is stored in Grbl's EEPROM, everytime the startup or reset is done; startup block printed back can be seen as a response from Grbl to indicate if it ran okay.

If there are multiple G-code startup blocks, it will be printed back in order upon every startup. And if one of the startup blocks needs to be removed, say block 0, it can be done by typing in $N0= without anything following the equal sign. Also, if the homing is enabled, the startup blocks will execute immediately after the homing cycle and not at startup.

$C - Check gcode mode

This toggles the Grbl's gcode parser to take all incoming blocks and process them completely, as it would in normal operation, but it does not move any of the axes, ignores dwells, and powers off the spindle and coolant. This is intended as a way to provide the user a way to check how their new G-code program fares with Grbl's parser and monitor for any errors.

When toggled off, Grbl will perform an automatic soft-reset (^X). This is for two purposes. It simplifies the code management a bit. But, it also prevents users from starting a job when their G-code modes are not what they think they are. A system reset always gives the user a fresh, consistent start.

$X - Kill alarm lock

Grbl's alarm mode is a state when something has gone critically wrong, such as a hard limit or an abort during a cycle, or if Grbl doesn't know its position. By default, if the homing is enabled and Arduino is powered up, Grbl enters the alarm state, because it does not know its position. The alarm mode will lock all G-code commands until the $H homing cycle has been performed. Or if a user needs to override the alarm lock to move their axes off their limit switches, for example, $X will kill alarm lock and override the locks and allow G-code functions to work again.

$H - Run homing cycle

This command is the only way to perform the homing cycle in Grbl. Some other motion controllers designate a special G-code command to run a homing cycle, but this is incorrect according to the G-code standards. Homing is a completely separate command handled by the controller.

After running a homing cycle, rather than jogging manually all the time to a position in the middle of the workspace volume, G28 or G30 pre-defined position can be set to be the post-homing position, closer to where it will be machining. To set these, machine needs to be jogged to location where it should move after homing. G28.1 (or G30.1) needs to be typed in to have Grbl store that position. After $H homing, G28 (or G30) needs to be entered and it'll move there auto-magically.

$RST=$, $RST=#, and $RST=*- Restore Grbl settings and data to defaults

These commands are not listed in the main Grbl $ help message, but are available to allow users to restore parts of or all of Grbl's EEPROM data. Grbl will automatically reset after executing one of these commands to ensure the system is initialized correctly.

- $RST=$: Erases and restores the $$ Grbl settings back to defaults, which is defined by the default settings file used when compiling Grbl. Often OEMs will build their Grbl firmware with their machine-specific recommended settings. This provides users and OEMs a quick way to get back to square-one, if something went awry or if a user wants to start over.
- $RST=#: Erases and zeros all G54-G59 work coordinate offsets and G28/30 positions stored in EEPROM. These are generally the values seen in the $# parameters printout. This provides an easy way to clear these without having to do it manually for each set with a G20 L2/20 or G28.1/30.1 command.
- $RST=*: This clears and restores all of the EEPROM data used by Grbl. This includes $$settings, $# parameters, $N startup lines, and $I build info string. This doesn't wipe the entire EEPROM, only the data areas Grbl uses.

Real-Time Commands: ~, !, ?, and Ctrl-X

The last four of Grbl's commands are real-time commands. This means that they can be sent at anytime, anywhere, and Grbl will immediately respond, no matter what it's doing. These are special characters that are picked-off from the incoming serial stream and will tell Grbl to execute them, usually within a few milliseconds.

~ - Cycle start

This is the cycle start or resume command that can be issued at any time, as it is a real-time command. When Grbl has motions queued in its buffer and is ready to go, the ~ cycle start command will start executing the buffer and Grbl will begin moving the axes. However, by default, auto-cycle start is enabled, so new users don't require this command unless a feed hold is performed. When a feed hold is executed, cycle start will resume the program. Cycle start will only be effective when there are motions in the buffer ready to go and will not work with any other process like homing.

! - Feed hold

The feed hold command will bring the active cycle to a stop via a controlled deceleration, so as not to lose position. It is also real-time and may be activated at any time. Once finished or paused, Grbl will wait until a cycle start command is issued to resume the program. Feed hold can only pause a cycle and will not affect homing or any other process. If it is needed to stop a cycle mid-program without losing position, a feed hold is performed to have Grbl bring everything to a controlled stop. Once finished, a reset can be issued.

? - Current status

The ? command immediately returns Grbl's active state and the real-time current position, both in machine coordinates and work coordinates. The ? command may be sent at any time and works asynchronously with all other processes that Grbl is doing. The active states Grbl can be in are: idle, run, hold, door, home, alarm and check.

- **Idle**: All systems are go, no motions queued, and it's ready for anything.
- **Run**: Indicates a cycle is running.
- **Hold**: A feed hold is in process of executing, or slowing down to a stop. After the hold is complete, Grbl will remain in Hold and wait for a cycle start to resume the program.
- **Door**: This compile-option causes Grbl to feed hold, shut-down the spindle and coolant, and wait until the door switch has been closed and the user has issued a cycle start. Useful for OEM that need safety doors.
- **Home**: In the middle of a homing cycle. Positions are not updated live during the homing cycle, but they'll be set to the home position once done.
- **Alarm**: This indicates something has gone wrong or Grbl doesn't know its position. This state locks out all G-code commands, but allows interaction with Grbl's settings if needed to. $X kills alarm lock and releases this state and puts Grbl in the idle state, which will let move things again.

- **Check**: Grbl is in check G-code mode. It will process and respond to all G-code commands, but not motion or turn on anything. Once toggled off with another $C command, Grbl will reset itself.

Ctrl-x - Reset Grbl

This is Grbl's soft reset command. It's real-time and can be sent at any time. As the name implies, it resets Grbl, but in a controlled way, retains the machine position, and all is done without powering down the Arduino. The only times a soft-reset could lose position is when problems arise and the steppers were killed while they were moving. If so, it will book if Grbl's tracking of the machine position has been lost. This is because an uncontrolled deceleration can lead to lost steps, and Grbl has no feedback to how much it lost. Otherwise, Grbl will just re-initialize, run the startup lines, and continue on its merry way.

It's recommended to do a soft-reset before starting a job. This guarantees that there aren't any G-code modes active that from playing around or setting up the machine before running the job. So, the machine will always start fresh and consistently, and it does what is expected.

Steps/mm calculations:-

Since the motors are not connected directly to lead screw for transmission of motion, the default value of $0 and $1needs to be changed based on calculation shown below.

Stepper steps/revolution (for stepper motor) = 200

Degree/step = 1.8^0

Total steps/revolution = 200

Current travel setting (Grbl)

Steps = 250

Millimeters of travel = 1

Millimeters of travel required for stepper to make one revolution

250 steps $\rightarrow$ 1mm

200 steps $\rightarrow$ x

x = 0.8mm

Actual distance travelled in 1 revolution

Timing pulley outer diameter = 9.56 mm

Number of revolution = 1

Pi = 3.1416

Distance = (Pi) (D) (Revolution)

= (3.1416) (9.56) (1)

= 30.033696 mm

Stepper steps per 1 mm

200 steps $\rightarrow$ 30.033696 mm

y steps $\rightarrow$ 1 mm

y = 6.659 steps/mm

This value obtained needs to be updated in Grbl settings for both x and y-axis. Just by entering $0=6.659 and $1=6.659 one at a time in the text box below the command option and hitting enter button, the values get updated. Verification of the above changes can be done by entering $$ in the same text box and hitting enter. The values can be seen getting reflected. Apart from the above two values default feed, default seed and acceleration values are also changed in similar fashion as shown in table 3.3.1.5. The values for these have been decided by taking trials on the machine.

Table 3.3.1.5: Comparison of default and changed settings

Symbol	Terminology	Unit	Default setting	Setting done to match hardware
$0	Travel across x-axis	step/mm	250	6.659
$1	Travel across y-axis	step/mm	250	6.659
$4	Default feed	mm/min	250	6000
$5	Default seek	mm/min	500	6000
$8	Acceleration	mm/sec^2	10	500

3.3.2. Machine control

Under this functionality, various axes can be reset to zero, soft reset can be achieved and machine can be made to return to zero. Unit selection functionality is provided under this option. Based on unit used for generating gcode, either inches or millimeters can be selected as shown in figure 3.3.2.1. Motion across various axes along positive and negative direction can be checked by respective buttons provided under this functionality.

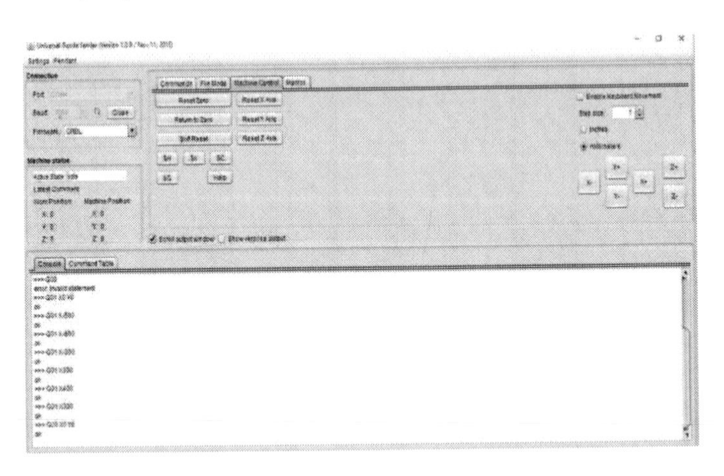

Figure 3.3.2.1: Machine control functionality

3.3.3. File mode

Under commands functionality we have seen we can input one line of gcode at a time. But what we want is to execute a group of commands or one complete operation or a group of operations in one go. This can be achieved by browsing the gcode file generated in UG NX 5 (discussed in Chapter 4) as shown in figure 3.3.3.1.

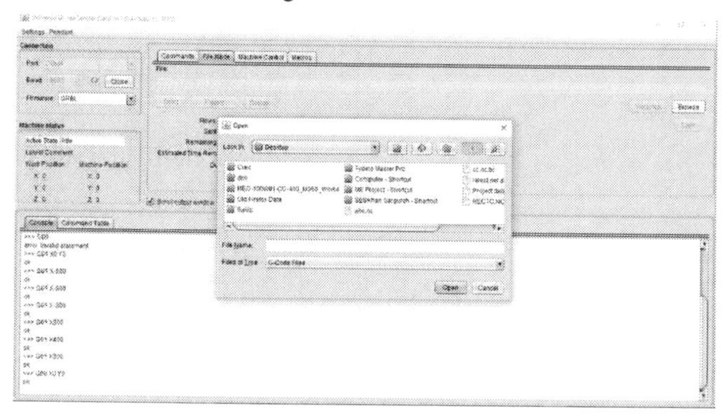

Figure 3.3.3.1: Browsing gcode file

After the file has been browsed, the image of the drawing to be drawn can be visualized by clicking on visualize button, next to browse button. The actual image of drawing along with tool can be seen as shown in figure 3.3.3.2.

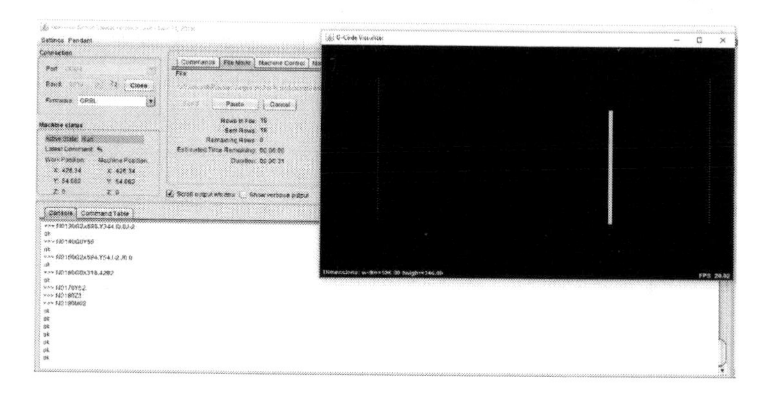

Figure 3.3.3.2: Gcode visualizer

On executing the gcode file by clicking on the send button, the motors get activated causing motion of mechanical members across x and y-axis to draw the desired image. The animated motion of the same can be simultaneously seen in the Gcode visualizer window. As the program is being executed, the status of the tool path can be seen under machine status option. Once the drawing operation is completed, a pop-up window indicating job completion appears.

Chapter 4

2D manufacturing of drawing file obtained from image file

4.1. Image to drawing file conversion

The JPG file of hexagon is first converted into a drawing file with extension .dxf using Img2CAD software. The software can be downloaded from the link http://www.img2cad.com/. Figure 4.1.1 shows the JPG image of a hexagon and Img2CAD application being used to convert it into drawing file having extension .dxf.

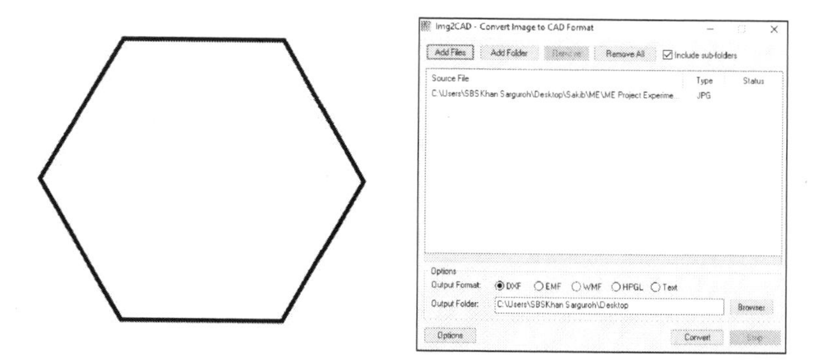

Figure 4.1.1: Hexagon JPEG Image and JPEG to CAD drawing converter

4.2. Importing drawing file into UG NX and scaling it

The obtained drawing file of hexagon is imported into UG NX 5.0 and is scaled to the required dimension as shown in figure 4.2.1. After conversion of JPG image into drawing file, the maximum dimensions across x and y axis were 9.5 and 9 mm respectively. Also, the Grbl controller has by default microstepping of 8x. Hence the scaling factor used is 48.8 and length across x and y axis of the hexagon are increased to 463.6 and 439.2 mm. Since the default microstepping setting 8x is used, the drawing obtained will be 57.95 and 54.9 mm across x and y axis of the hexagon. After conversion of the image file to drawing file, there are variations across lengths of sides of hexagon between 4.6 mm to 4.8 mm.

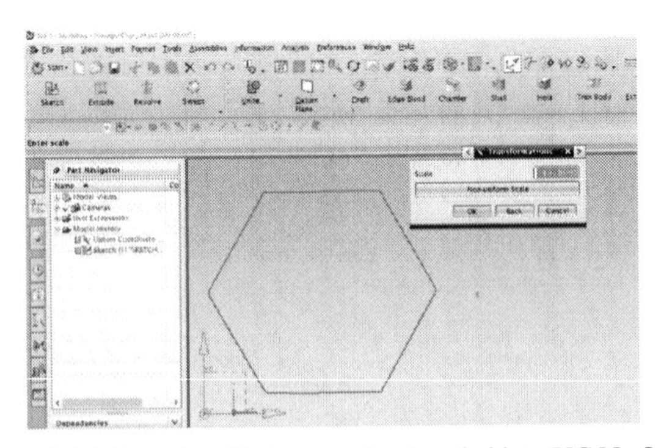

Figure 4.2.1: Drawing file imported and scaled into UG Nx 5.0

4.3. 2D manufacturing and gcode generation

2D manufacturing is carried out onto the scaled drawing and post-processing is done to obtain G and M codes of the required drawing which is saved with extension .nc. This section gives detailed explanation on how 2D manufacturing is carried out to obtain G and M code file in UG NX 5.

4.3.1. Initializing machining environment

The machining environment to be selected is mill planar which is done by first clicking on manufacturing application under start navigation. A machining dialog box appears in which mill planar option is selected as CAM setup by clicking on initialize option.

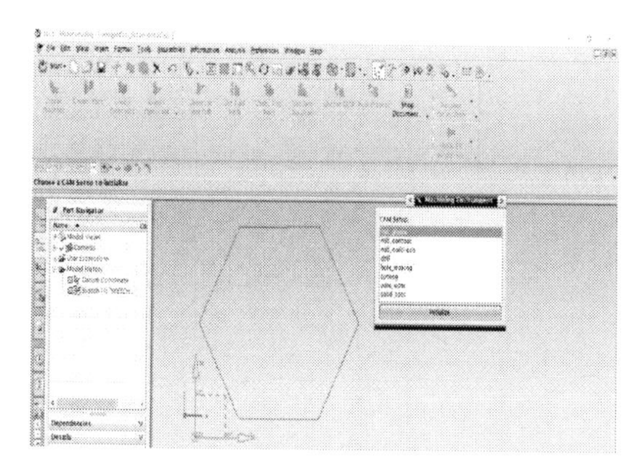

Figure 4.3.1.1: Initializing CAM setup as Mill planar

4.3.2. Machine coordinate system

By right clicking on operation navigator, geometry view is selected. Then MCS_Mill option is double clicked and machine coordinate system dialog box appears as shown in figure 4.3.2.1. Under clearance option, plane option is selected and icon besides plane is clicked for plane constructor dialog box to appear as shown in figure 4.3.2.2. Under plane constructor dialog box offset is kept zero and XY plane is selected.

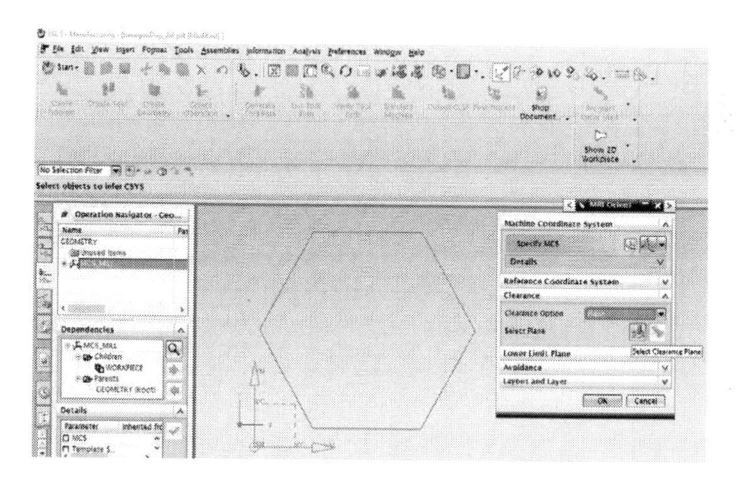

Figure 4.3.2.1: Machine coordinate system

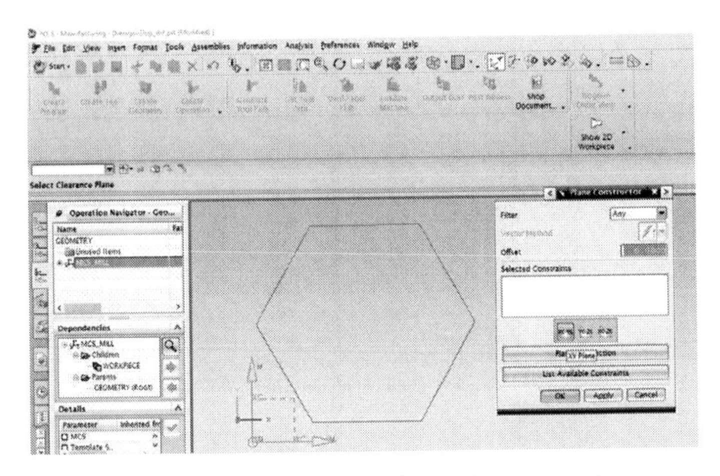

Figure 4.3.2.2: XY plane selection in Plane constructor

4.3.3. Machine tool creation

By right clicking on MCS mill option machine tool view can be selected and on the ribbon above create tool option appears. By clicking on this option create tool dialog box appears. Mill icon is selected in subtype as shown in figure 4.3.3.1. After clicking on OK milling tool parameters dialog box appears in which diameter of tool is set as 5 mm as shown in figure 4.3.3.2.

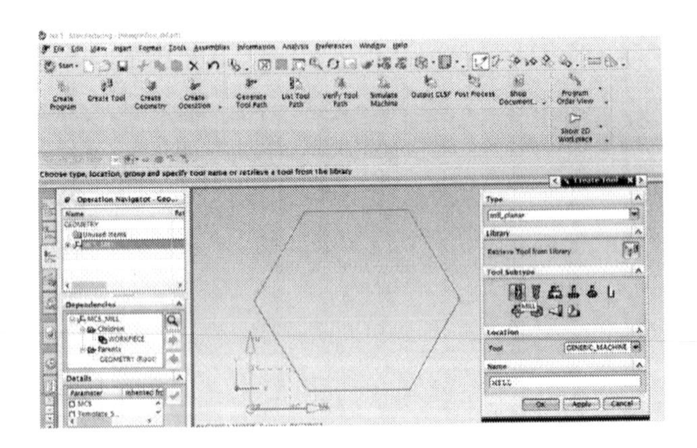

Figure 4.3.3.1: Mill tool creation

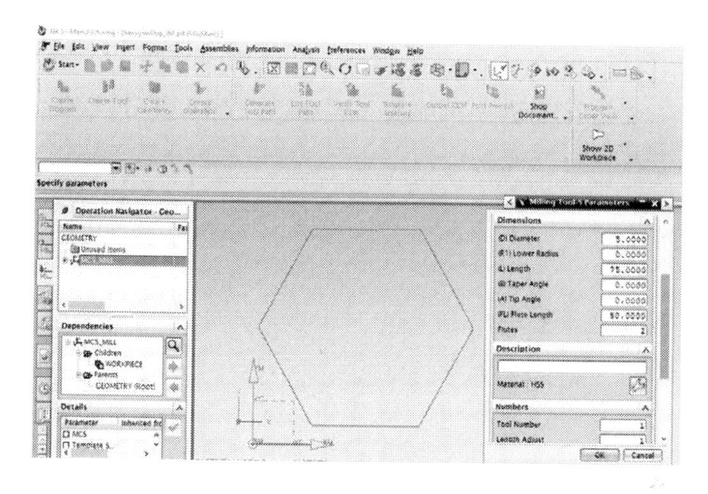

Figure 4.3.3.2: Setting diameter in milling tool parameter

4.3.4. Machine tool operation creation

After right clicking on MCS mill, operation option is selected and create operation dialog box appears. Planar operation is selected as operation subtype as shown in figure 4.3.4.1. Under location option all options are kept default but method option is changed to Mill_Finish.

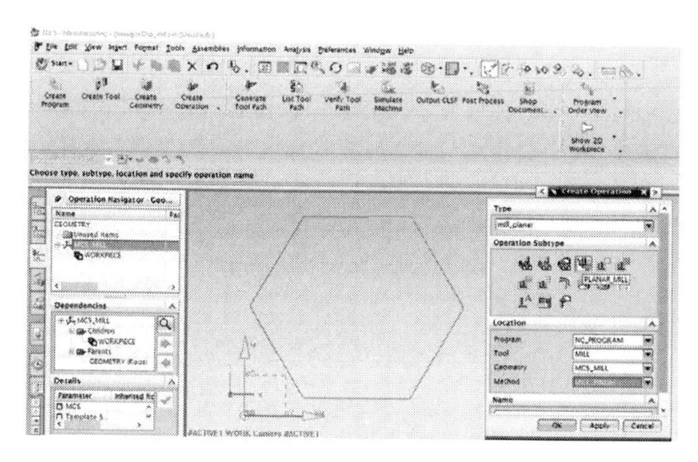

Figure 4.3.4.1: Planar mill operation creation

4.3.5. Specifying part and floor geometry

Under geometry option of planar mill dialog box as shown in figure 4.3.5.1, icon of specify part boundaries is clicked.

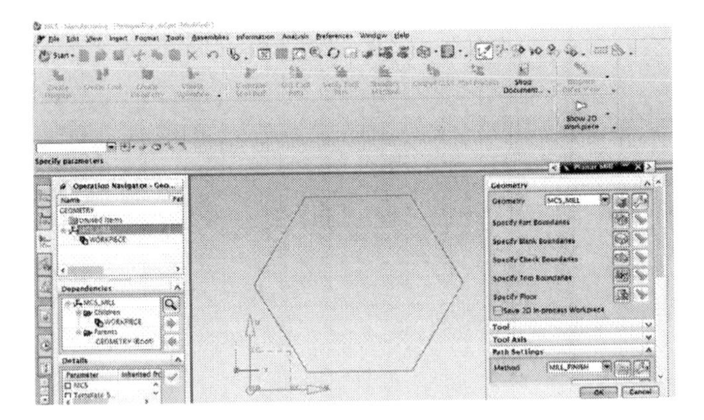

Figure 4.3.5.1: Planar mill settings

After clicking on specify part boundaries icon, boundary geometry icon appears in which curve/edges is selected under mode drop down option as shown in figure 4.3.5.2. Under create boundary dialog box closed option is selected under type drop down menu as shown in figure 4.3.5.3. Then hexagon drawing is selected as the geometry.

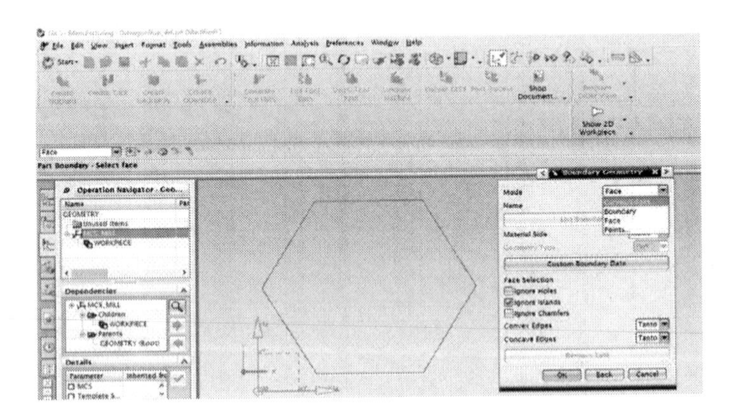

Figure 4.3.5.2: Boundary geometry mode selection

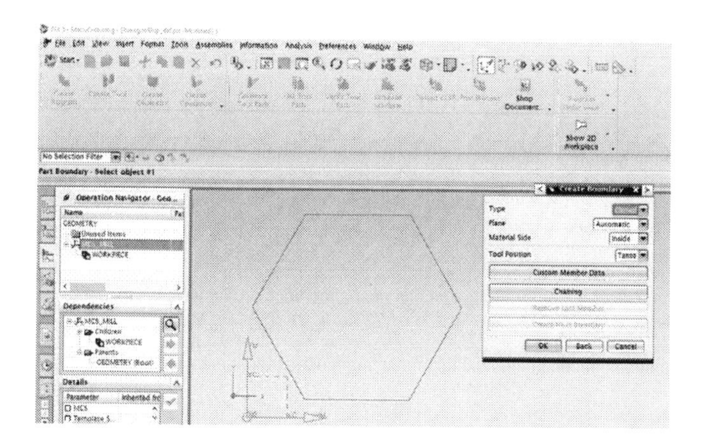

Figure 4.3.5.3: Boundary type selection

After clicking on the specify floor option icon, plane constructor dialog box appears in which offset is kept as zero and XY plane is selected as shown in figure 4.3.5.4.

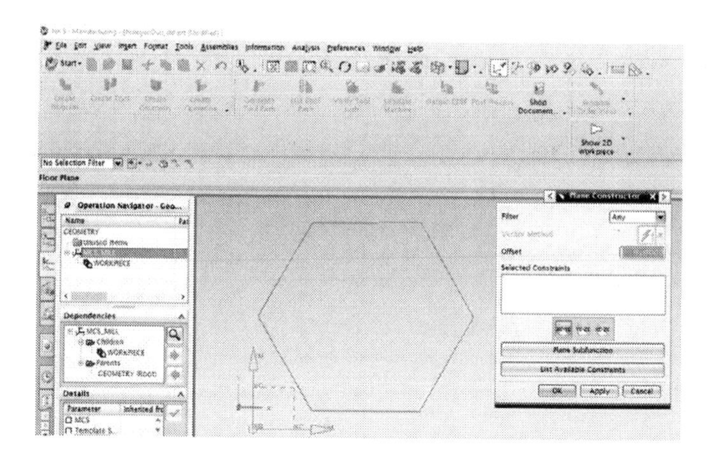

Figure 4.3.5.4: XY plane selection in plane constructor

4.3.6. Tool path settings

Various path settings associated with tool is available under path settings option. Under cut pattern, follow periphery option is selected as shown in figure 4.3.6.1.

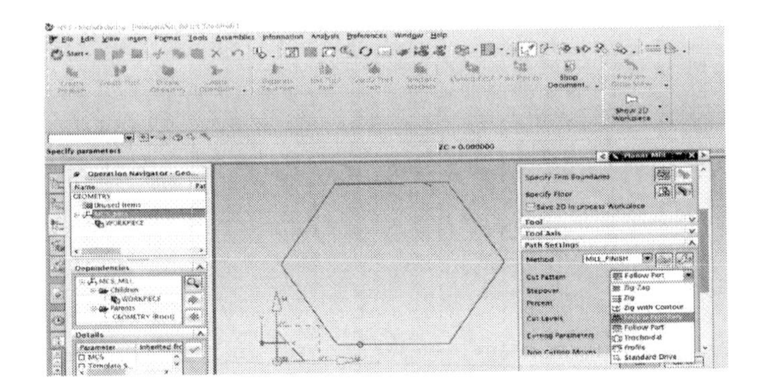

Figure 4.3.6.1: Cut pattern

Cutting parameters option icon is clicked and add finish passes option is clicked in the strategy tab as shown in figure 4.3.6.2.

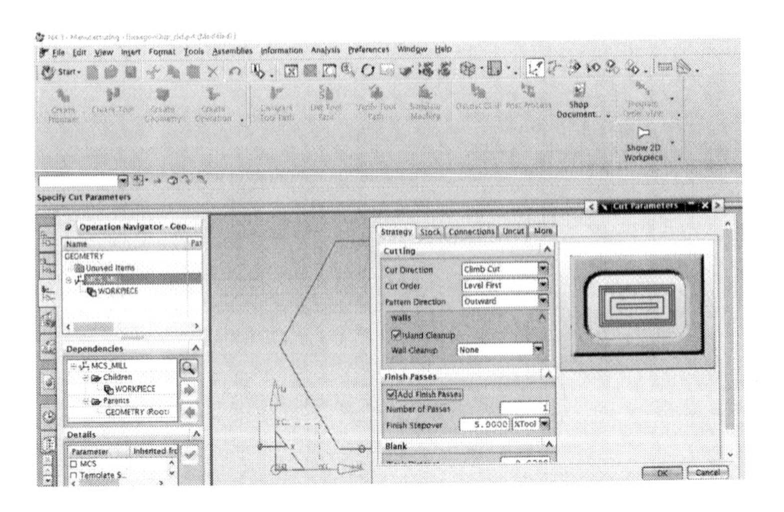

Figure 4.3.6.2: Cut parameter

Feed rate is set to 6000 mmpm under feed dialog box as shown in figure 4.3.6.3. This value is fed based on the value set in Grbl setting using universal gcode sender software application.

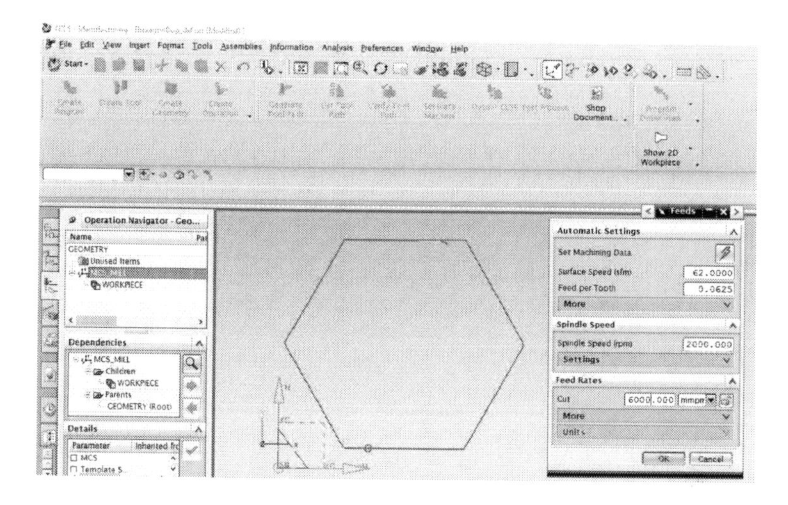

Figure 4.3.6.3: Feed rate

4.3.7. Tool path generation, verification and visualization

After the desired parameters have been set, generate icon is clicked under Actions option. This causes generation of tool path along hexagon boundary which is indicated by blue color as shown in figure 4.3.7.1.

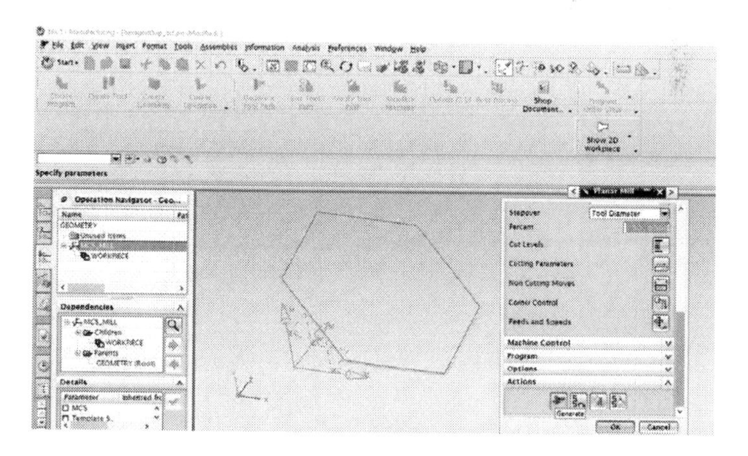

Figure 4.3.7.1: Tool path generation

Then by clicking on verify option, tool visualization dialog box appears through which actual motion of the tool path across the geometry i.e. hexagon boundary can be seen in animated form as shown in figure 4.3.7.2.

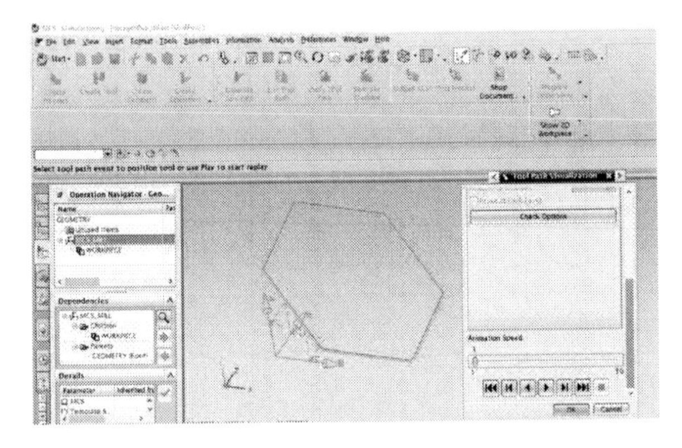

Figure 4.3.7.2: Tool path visualization

4.3.8. Post processing

After successful generation of tool path, post processing of the same is done by clicking on post process option on the ribbon bar above the drawing area. Amongst the various post processor options available, Mill 3 axis option is selected as shown in figure 4.3.8.1. Under Units option, Metric/Part is selected.

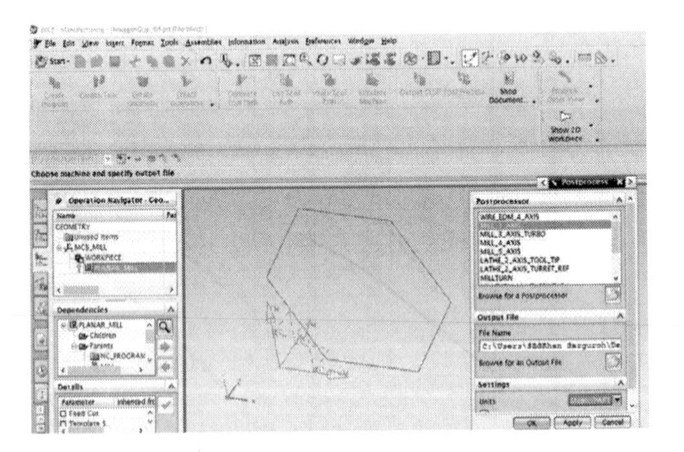

Figure 4.3.8.1: Post processor and unit selection

After clicking on OK, a gcode file is generated which is saved with extension .nc as shown in figure 4.3.8.2. This file is taken as input by Universal GcodeSender software to run the steppers motors incorporated to a mechanical system and draw the desired drawing of the hexagon. Similar gcode generation process is carried out for rectangle and circle.

Information

File Edit

```
%
N0010 G40 G17 G90 G70
N0020 G91 G28 Z0.0
:0030 T01 M06
N0040 G0 G90 X316.3247 Y455.5522 S2000 M03
N0050 G43 Z3. H01
N0060 Z0.0
N0070 X316.342 Y450.5523
N0080 X361.1028 Y450.7066
N0090 G2 X365.4458 Y448.2142 I.0172 J-4.9999 F600.
N0100 G0 X401.267 Y386.4177
N0110 X437.0238 Y324.5794
N0120 X480.1325 Y249.7492
N0130 G2 X480.7497 Y247.9605 I-4.3325 J-2.4959
N0140 G0 X481.5631 Y242.2671
N0150 G2 X480.944 Y239.0609 I-4.9498 J-.7071
N0160 G0 X461.0029 Y204.5051
N0170 X441.0946 Y169.9344
N0180 X365.4563 Y38.1773
N0190 G2 X361.1417 Y35.6667 I-4.3363 J2.4894
N0200 G0 X174.0751 Y34.8534
N0210 G2 X171.3199 Y35.6667 I-.0218 J4.9999
N0220 G0 X127.6933
N0230 G2 X124.1578 Y37.1311 I0.0 J5.
N0240 G0 X123.7707 Y37.5182
N0250 G2 X120.6576 Y40.3237 I1.4826 J4.7751
N0260 G0 X119.1274 Y43.8942
N0270 G2 X117.6539 Y45.5145 I2.8726 J4.0925
N0280 G0 X68.2957 Y132.2895
N0290 G2 X66.4297 Y134.1404 I2.4643 J4.3505
N0300 G0 X8.43 Y234.6187
N0310 X5.8299 Y242.4189
N0320 G2 X6.1616 Y246.3529 I4.7435 J1.5811
N0330 G0 X12.7076 Y258.6268
N0340 X94.8813 Y401.0069
N0350 X120.8934 Y447.341
N0360 G2 X125.2361 Y449.8933 I4.3599 J-2.4477
N0370 G0 X316.342 Y450.5523
N0380 X316.3247 Y455.5522
N0390 Z3.
N0400 M02
%
```

Figure 4.3.8.2: Gcode file generated

Chapter 5

Experimental Setup

The mechanical system in the project comprises of four aluminium angle brackets (two for each axis), four chromed rods which are fitted on brackets along which linear ball bearings slide. These linear ball bearings move across x and y axis by stepper motor via timing belt and pulley to draw the desired figure. This section gives an insight into various mechanical components used in the project.

5.1. Aluminium angle brackets

These form the base of the experimental setup on which various other parts are assembled. Four angle brackets are used, two for each x and y axis respectively as shown in the figure 5.1.1.

Figure 5.1.1: Aluminium angle brackets

Holes are drilled across various points on angle brackets to fit 8 mm diameter chromed rods and stepper motors.

5.2. Chromed rods

8 mm diameter chromed rods of 500 and 200 mm length are fitted across two sets of angle brackets for x and y axis respectively as shown in figure 5.2.1. Linear bearings slide along these rods.

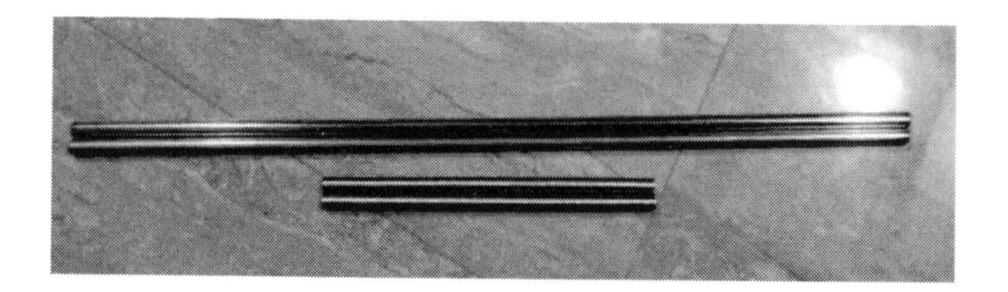

Figure 5.2.1: Chromed rods

5.3. Linear motion bearings

SC8UU linear motion bearings are used which slide along the chromed rods. Two sets are used for each axis i.e. four linear motion bearings are used as shown in figure 5.3.1.

Figure 5.3.1: SC8UU linear motion bearings

Linear bearings sliding across x-axis i.e. 500 mm chromed rods are fitted to the y-axis angle brackets using screws enabling complete motion of y-axis setup across x-axis. Linear bearings sliding across y-axis i.e. 200 mm chromed rods are fitted to an Aluminium block to which a pencil or sketch pen will be fitted for drawing.

5.4. Aluminium block to fit pencil for drawing

A 25x50x50 mm aluminium block is used in which pencil or sketch pen is fitted for drawing the desired image as shown in figure 5.4.1.

Figure 5.4.1: Block for holding pencil or sketch pen

The pencil is passed through the hole drilled on top face of the block which is tightened through a screw passed through threaded hole along the side face of the block. This block is fitted to two linear motion bearings which are passing through 200 mm chromed rods across y-axis. As linear bearings across x and y-axis move, the pencil or sketch pen fitted in the block will draw the desired drawing.

5.5. Stepper motor

2.6 kg-cm bipolar stepper motors shown in figure 5.5.1 are used which receive pulses from drivers on Grbl shield V5 causing it rotate in required amount of steps. Two stepper motors have been used; one fitted on one pair of aluminium angle brackets for obtaining motion across x-axis; the second is fitted on the other pair of angle brackets for obtaining motion across y-axis.

Figure 5.5.1: 2.6 kg-cm bipolar stepper motor

These motors are controlled by Grbl shield enabling simultaneous motion across x and y-axis. This circular motion in steps is converted into linear motion of bearings by fitting a timing pulley over shaft of motor. Motor motion is transmitted to timing pulley which further causes linear motion of timing belt causing motion of x and y mechanical systems.

5.6. Timing pulley

Timing pulley used is GT2 16 teeth 5mm bore as shown in figure 5.6.1. It is fitted over the stepper motor shaft by means of two grub screws provided in its two threaded holes.

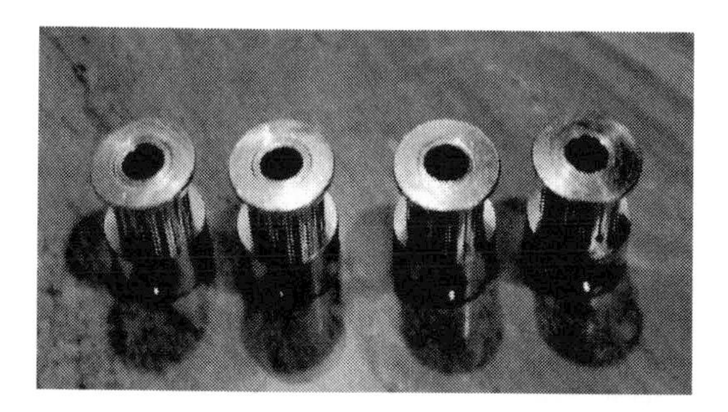

Figure 5.6.1: GT2 16 teeth 5mm bore timing pulley

For producing motion across one axis two timing pulleys are used; one fitted over the shaft of motor and other fitted at opposite end over the aluminium angle bracket by means of 10-24 socket head cap screw.

5.7. Timing belt

Timing belt used is GT2 open timing belt 6mm width as shown in figure 5.7.1. For motion across x-axis it is wrapped across two pulleys (one pulley mounted over shaft and other mounted over opposite angle bracket) with its ends kept open. The open ends of the belt are fitted to the threaded holes drilled in one of the y-axis angle bracket by means of screw. For motion across y-axis, timing belt is wrapped in similar fashion as that of x-axis but its open ends are fitted in threaded hole of 25x50x50 mm aluminium block which is holding a pencil for drawing purpose.

Figure 5.7.1: GT2 timing belt 6mm width

As shaft of motor rotates, motion is transmitted to timing belt via timing pulley, causing complete y-axis setup motion across x-axis and motion of block holding pencil or sketch pen across y-axis.

5.8. Complete assembly

Apart from the mechanical components discussed in this chapter; the electrical, electronic and software components discussed in chapter 3 also form part of assembly as shown in figure 5.8.1.

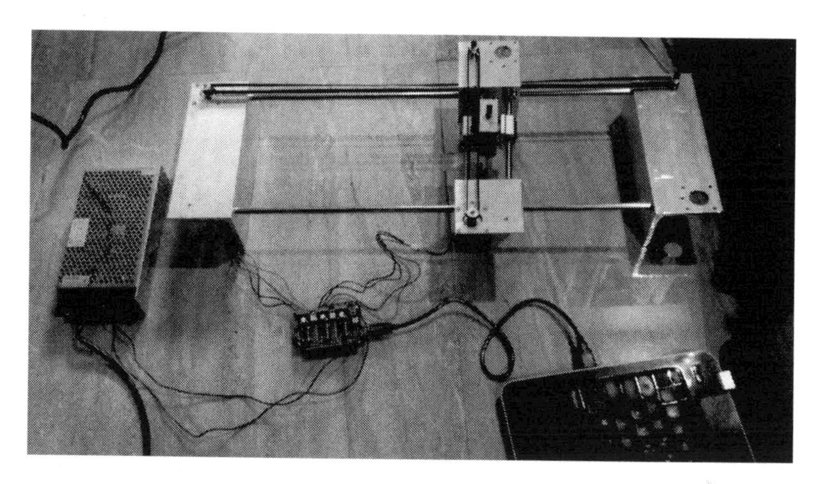

Figure 5.8.1: Experimental setup

5.8.1. Executing the gcode program and running the setup

The gcode file generation of hexagon shown in chapter 4 is given as input to Universal GcodeSender software (discussed in chapter 3). After clicking on the visualize button, the hexagon to be drawn along with the tool path can be seen as shown in figure 5.8.1.1. Similar visualization can be seen before executing gcode program for a circle as shown in figure 5.8.1.2.

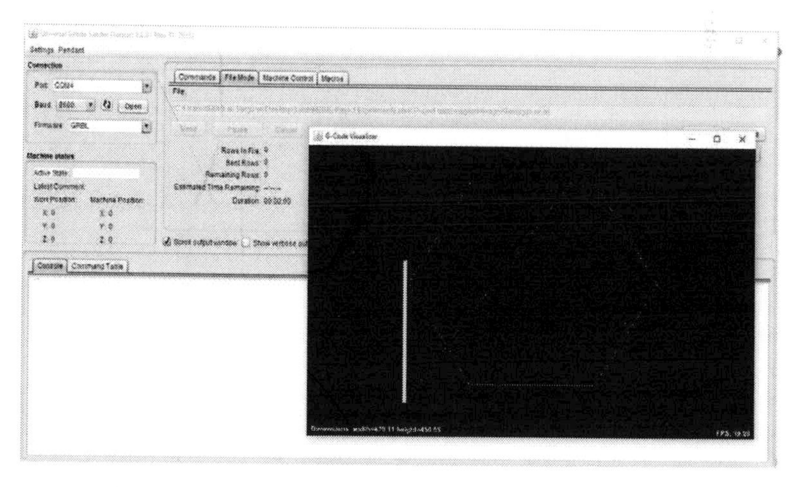

Figure 5.8.1.1: Gcode visualizer depicting tool path for hexagon

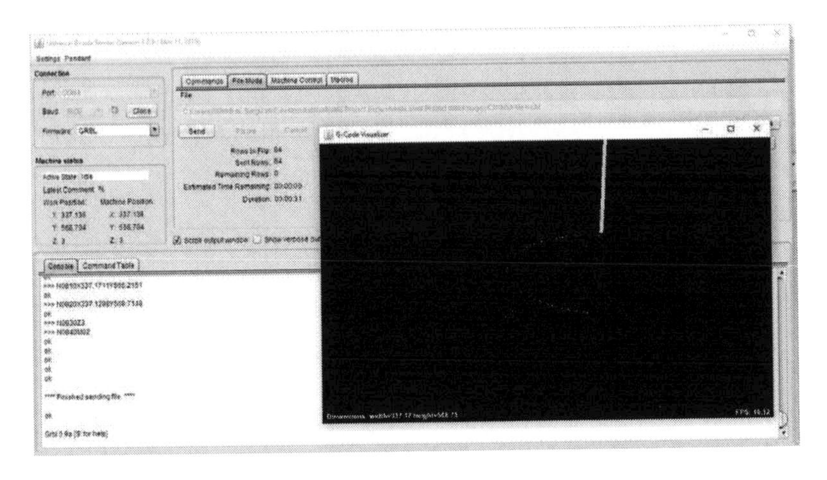

Figure 5.8.1.2: Gcode visualizer depicting tool path for circle

After executing the gcode program by clicking on send button, the stepper motors are actuated causing motion of block holding pencil or sketch pen across x and y axis via timing pulley, timing belt and other mechanical parts. Figure 5.8.1.3 shows hexagon and figure 5.8.1.4 shows circle drawn using the setup.

Figure 5.8.1.3: Hexagon drawing using setup

Figure 5.8.1.4: Circle drawing using setup

Chapter 6

Results and discussion

Based on the experiment performed, the following observations have been made:

1) Variation of 4.1% in sides of hexagon has been observed after converting the image file into drawing file affecting the overall accuracy by the same variation.

2) Variation of 5.3% has been observed in diameter across x and y axis after converting the image file of circle into drawing affecting the overall accuracy by same variation.

3) Small arcs are formed at corners of hexagon sides causing reduction in speed of machine when it changes path.

Table 6.1 gives a comparison between various parameters of hexagon and circle drawn in modelling software UG NX 5 to that drawn by the two axis CNC machine (discussed in Chapter 5).

Table 6.1: Comparison of dimension in Nx and drawn through machine

	Hexagon			**Circle**
	Parameters (in mm)			**Parameters (in mm)**
	Maximum length across x-axis	**Maximum length across y-axis**	**Length of side**	**Diameter**
Dimension in UG NX 5	57.95	54.9	29.3	60.1
Dimension obtained on machine	56.1	52.7	30	61

Chapter 7

Conclusion and future scope

Proprietary control of CNC machines results in inability to control and enhance inputs, since the program codes cannot be modified as the vendors distribute them in compiled form. In this book, a method has been presented in eliminating the proprietary control by utilizing a universal Grbl-Arduino based controller which takes in gcode generated by any CAD/CAM software. The gcode program is then executed by means of windows based application, Universal GcodeSender, to draw the desired figure using a simple two axis hardware setup. Following objectives have been achieved during the course of the project:

1) A universal controller was successfully developed which takes G and M codes generated by UG NX 5.0 CAD/CAM software.

2) A JPEG or PNG image of a rectangle or circle or polygon was successfully converted into drawing file (.prt or .dxf) and gcode file for the same was generated by carrying out 2D manufacturing in UG Nx 5.0.

3) A simple two axis CNC machine was successfully developed which was used to draw the desired rectangle or circle or polygon after getting respective gcode input through Universal GcodeSender software.

4) Accuracy of the drawn circle and hexagon was measured and there were slight variations in the range 2 to 5%.

This work can be further continued to develop a three axis CNC machine as Grbl shield V5 has three inbuilt stepper drivers out of which only two have been used for a stepper motor each in the current project. Using the third driver will facilitate usage of third axis causing the machine to work as a milling machine. Hardware for third axis will be required to be further developed for incorporating z axis motor.

Chapter 8

References

1) Banzi Massimo, Cuartielles David, Mellis David, Igoe Tom, Martino Gianluca. Arduino Uno Hardware. Retrieved from https://www.arduino.cc/en/Main/ArduinoBoardUno on 2015, June 4.

2) Banzi Massimo, Cuartielles David, Mellis David, Igoe Tom, Martino Gianluca. Arduino Uno Software. Retrieved from https://www.arduino.cc/en/Main/Software on 2015, June 5.

3) Banzi Massimo, Shiloh Michael. Getting started with Arduino, 3rd Edition. Retrieved from http://shop.oreilly.com/product/0636920029267.do on 2015, June 11.

4) Chengrui Zhang, Heng Wang, Jingkun Wang, August 2003, "An USB-based software CNC system," Journal of Materials Processing Technology, Volume 139, Issues 1-3, pp. 286-290.

5) D. Dumura, P. Bouchera, J. Röderb, 1998, "Advantages of an Open Architecture Structure for the Design of Predictive Controllers for Motor Drives," CIRP Annals - Manufacturing Technology, Volume 47, Issue 1, pp. 291-294.

6) Evans Brian, 2012, Practical 3D Printers, Apress, New York.

7) G. Pritschowa, C. Kramera, 2005, "Open system architecture for drives," CIRP Annals - Manufacturing Technology, Volume 54, Issue 1, pp. 375-378.

8) G. Pritschow, Ch. Daniel, G. Junghans, W. Sperling, 1993, "Open System Controllers - A Challenge for the Future of the Machine Tool Industry," CIRP Annals - Manufacturing Technology, Volume 42, Issue 1, pp. 449-452.

9) Günter Pritschow (Co-ordinator), Yusuf Altintas, Francesco Jovane, Yoram Koren, Mamoru Mitsuishi, Shozo Takata, Hendrik van Brussel, Manfred Weck, Kazuo Yamazaki, 2001, "Open controller architecture-Past, present and future," CIRP Annals - Manufacturing Technology, Volume 50, Issue 2, pp. 463-470.

10) Kazuo Yamazakia, Yoshimaro Hanakib, Masahiko Moric, Kazusaka Tezukad, 1997, "Autonomously Proficient CNC Controller for High-Performance Machine Tools Based on an Open Architecture Concept," CIRP Annals - Manufacturing Technology, Volume 46, Issue 1, pp. 275–278.

11) Saunders John. DIY Arduino CNC machine with GRBL Shield. Retrieved from https://www.youtube.com/watch?v=1ioctbN9JV8 on 2015, July 10.

12) Skogsrud Simen Svale, Jeon Sungeun K. Grbl. Retrieved from https://github.com/grbl/grbl/wiki on 2015, August 16.

13) Skogsrud Simen Svale, Jeon Sungeun K. Using grblShield. Retrieved from https://github.com/synthetos/grblShield/wiki/Using-grblShield on 2015, September 4.

14) Stephen J. Rober, Yung C. Shin, June 1995, "Modeling and control of CNC machines using a PC-based open architecture controller," Mechatronics, Volume 5, Issue 4, pp. 401-420.

15) S.T. Newman, A. Nassehi, 2007, "Universal Manufacturing Platform for CNC Machining," CIRP Annals - Manufacturing Technology, Volume 56, Issue 1, pp. 459-462.

16) Xiong-bo MA, Zhen-yu HAN, Yong-zhang WANG, Hong-ya FU, June 2007, "Development of a PC-based Open Architecture Software-CNC System," Chinese Journal of Aeronautics, Volume 20, Issue 3, pp. 272-281.

17) X. Wei, C. Jihong, Y. Jin, 2009, "Design of Servo System for 3-Axis CNC Drilling Machine Based on xPC Target," IEEE International Colloquium on Computing, Communication, Control, and Management, pp. 447-450.

18) Yusri Yusofa, Kamran Latif, 2015, "New Interpretation Module for Open Architecture Control based CNC Systems," Procedia CIRP, Volume 26, pp. 729-734.

8.1. Publications

1) Sakib S. Sarguroh, Arun B. Rane. Using GRBL-Arduino-based controller to run a two-axis computerized numerical control machine. IEEE Xplore, DOI: 10.1109/ICSCET.2018.8537315,19th November, 2018.

2) Sakib S. Sarguroh, Arun B. Rane, Saurabh A. Korgaonkar, D. S. S. Sudhakar. Elimination of proprietary control for Computerized Numerical Control (CNC) Machine. Journal of Basic and Applied Research International, ISSN: 2395-3438 (P), ISSN: 2395-3446 (O), Vol.17, Issue 3, pp. no. 211-217, 28th April, 2016.

Printed in Great Britain
by Amazon

68421563R00045